WILD LIFE

A GARDEN DRAMA

Conceived, designed and written by the Green Queen

Wild Life
A Garden Drama

First published in Great Britain in 2005 by Green Queen Books

Pictures:
Andy Anderson (celandines)
Lynne Carlisle (Green Queen)
Paul Hobson (fox, badger, frog, butterfly)
Mike Lane (blackberries, thrush, squirrel, dragonfly, blackbird, feeder, spider's web, toadstool and frog)
Other pictures kindly given by Fantasia, Demonike, Taut, Kel, etc

Cover design: Nicola Harrison

Printed and bound by DS Publishing (Sheffield)

Contents

The Green Queen feeding tadpoles

The Scene

The Green Queen

The Green Queen grew up in the wild woods. In an old fort with high walls and a big hill in the middle. It was ramshackle. A ruin. The Green Queen lived there with her grandmother, in an old lean-to stuck on the inside of the main wall.

The Green Queen's tiny bedroom had ferns growing up the walls and a clear plastic roof over her bed. Stripy snails crawled across her ceiling. Frogs croaked from under her bed, and at night she would lie awake and gaze at the stars.

Outside the lean-to, there was a tangled wilderness of brambles, ivy, and wildflowers. The nettles were five feet tall and covered with caterpillars. The undergrowth was lush and the air vibrated with birdsong.

This is where the Green Queen began - right in the heart of nature. It is also where the Green Queen project began, because to know nature is to love it forever.

Introduction

This quirky little book was created as part of the Green Queen Project, an initiative set up four years ago in response to growing concerns about the environment. Amidst plummeting numbers of many species and dire predictions for the future of our natural world, the Green Queen was born.

Some believe she is a tree spirit. Others say that she is a fairy with a human face. And the rest say nothing at all because they are too busy spraying their roses with chemicals.

The Green Queen Project began in South Yorkshire and then grew a *lot.* There were three main phases to the project, centring around public information, journalism, education in schools, talks and lectures.

This book is the sum of all that work and experience. Like every other aspect of the project, it was written free of charge. Printing and publishing costs were met by the sale of a Georgian cabinet, an oboe, and forty pairs of second-hand shoes.

Acknowledgements

Thank you to everyone who donated time and money to this wonderful project.

To all the garden centres and florists who kindly sold our Green Queen wildflower seeds without taking a penny themselves.

To those newspapers and radio stations that gave us masses of coverage. And to Alan Powell, Editor of the Sheffield Telegraph, for his support of the project.

Also to Lynne Carlisle, Helen Vaughan and Lynne Middleton for their commitment, support and occasional leaps of faith.

Finally to the Green Queen - a real person, as it happens, and quite a wild one - who was the inspiration and driving force behind the whole thing.

To the reader

Here is my little book. Love it and read it well for it is a thing of the heart, and like all things of the heart, it can only do good in the world.

It does not seek to be comprehensive or scientific. It is a simple, practical guide written in plain English for everyone to read. There is no cleverness or showy prose. This is a book with a message, and its message is all that matters.

But it is also an invitation. An invitation to think differently about the world around you, about the part you play in that world, and your responsibility for it. For our natural world is beautiful, and like all beautiful things, it is delicate and fragile. We need to make it more robust. To strengthen the links that hold it together.

What this book shows is that a few simple but informed changes in our gardens can make a huge difference to a world where the survival of so many species is now in the balance.

Information is the key to change. So please read my book.

Green Queen

Green Queen

Walk on the Wild Side

The joy of a wild garden is that it has a real purpose. Everything is grown and tended for a particular reason, creating a harmonious, useful space.

It's a different way of gardening, a different way of seeing - and the result is a different kind of beauty. The wildlife garden is all curves and hidden places, birds peeping out of tangled ivy, butterflies dancing from scabious to elder. There is the sound of trickling water, the soft hum of countless bees.

Gardening for the wild is not an excuse to neglect your garden. It's an invitation to plan and think about it differently - to use every inch of it, and to make every little corner count. The wildlife garden is not left to nature. It is actively given over to it.

SO GIVE WILDLIFE A CHANCE

Flower Power

Choosing plants for the garden is a matter of taste. Choosing plants for wildlife is a matter of tasting good. Your average bird or bee isn't interested in a bright show of annuals. What they like are weeds and seeds. They want grub. Or rather, grubs.

Ease up on the weed. Nettles, thistles and docks may not be the greatest lookers, but to a passing bird they are simply stunning. As well as providing food for caterpillars and insects, weeds are prolific seed producers. Chickweed, groundsel, plantain and dandelion are adored by finches, while grassy seed-heads are much beloved of sparrows. So go on, have a wild patch. Fill it with grasses, tall buttercups, poppies, cornflowers, campion, oxeye daisies, teasels.

Herbicidal tendencies. Never spray weedkiller or insecticide (see 3 The Plot/Chemical Warfare for alternatives).

Social climbers. Hang out the honeysuckle, clutter with clematis. And thicken your hedges with guelder and dog rose - they add hips, prickles and dense cover. Don't forget the ivy, either - wonderful for nesting, good berries for birds, and home to many insects and spiders.

Bee happy. Butterflies like nectar-rich flowers; buddleia, knapweed, thistle, scabious, cornflower, viper's bugloss, yarrow, meadowsweet, ivy flower. Bees like pollen; foxgloves, hollyhocks, snapdragons, clover, lavender, borage, most herbs, evening primrose. And they're mad about sedum (see 3 The Plot/Things with Wings).

Fruit case. Soft fruiting plants such as gooseberries, blackberries and wild strawberries can be left over for birds and hedgehogs.

Back to the bush. Get in the right trees and shrubs - the pricklier the better! Holly is brilliant for nesting and feeding. Elder is a complete star - fast-growing, prolific umbrels of flowers, and berries to feed the five thousand. Hawthorn is the leading lady, same reasons as above, leaves eaten by moth larvae. Blackthorn is good (finches like the sloes), and rowan berries attract fieldfares and waxwings. Don't forget cotoneaster and japonica, either.

Dead wood. Old logs are host to thousands of minibeasts, particularly beetles, and brilliant hibernation spots, too. They also make a feature of dark corners where nothing much will grow. So make a pile and let it rot.

Shady characters. There are some wildflowers that grow under trees and hedges, as well as native spring bulbs (bluebells, snowdrops, wild daffodils, aconites and wood anemones) and early Spring wildflowers like celandine. Ferns are beautiful and thrive in shade, while species

like foxglove, herb Robert, red campion, primrose, violet, comfrey and wild strawberry will also do well.

Get nettled. Great food for butterfly caterpillars - and for the birds that feed on them. A double bonus! If you're concerned about nettles becoming invasive, plant some in a large tub to contain the roots.

Go native. Home-grown is always best. British wildflowers and trees are beautiful, and tailor-made for British wildlife. They are also hardy, suited to our soil and climate and will thrive without much attention.

High five. Leave 5 sq metres of lawn to grow long at the back, giving a lovely grassy corridor for creatures.

Decked out. Avoid decking, pebbling, extended patios and tarmac. They're the scourge of wildlife - and they're uneasy on the eye.

The Characters

1. The Goodies

Hog in the Headlights

They don't call him a hog for nothing. He snorts at the table, talks with his mouth full and tramples on his own food. His manners are foul *and* he's got fleas. Yet Spike is irresistible. He's been on our planet for 15 million years and we love to see him snuffling along the hedgerows. We love him to death. Literally. Every year we kill around 75,000 hogs on our roads.

Hedgehogs are nocturnal and will travel up to two kilometres a night, often crossing many roads in search of food. In a safe world, hedgehogs live up to ten years. In our world they rarely live beyond 2-3 years.

Apart from road kill, the other main claimant of hedgehog life is hibernation. Many hogs succumb in their first year because they have not stacked up enough body weight (over 500g) to see them through the winter. This is why they must eat, eat, eat.

HELPING HOGS IN THE GARDEN

- Hedgehogs eat beetles, caterpillars, slugs and snails so don't use pesticides and other chemicals if you want to help them. Organo phosphates are particularly dangerous to hogs.

- In the garden, build ponds with sloping sides or with stone ramps so that clumsy hogs that have fallen in can get out again.

- Pick up litter, empty tins, plastic carriers, garden twine and any netting that might be lying around as these can become deathly snares for hedgehogs. If you *must* use netting, try a fixed chicken wire enclosure in place of netting, or use hedgehog-safe netting (*Netlon* from Homebase).

- Hundreds of hogs burn every year because they crawl into piled bonfires to hibernate. Always check through carefully before lighting - or use an incinerator instead.

- Cattle grids are dangerous. Passing hedgehogs fall in but cannot get back out. Sadly, they will die of starvation. Prevent this by building a ramp of stones at one side of the grid so that hogs can clamber out again.

- Compost heaps are much beloved of hedgehogs. So please don't prod around yours with a pitchfork.

- Keep garage, shed and outhouse doors shut to stop hogs and other creatures wandering in and becoming trapped and starving to death. And always check before closing up.

- Hogs like to snooze in the longer grass at lawn edges, so beware when strimming and mowing. A maimed hedgehog is a terrible sight.

- NEVER GIVE HOGS MILK. It makes them very ill. In the evening, put out a small plate of cat or dog food (non-fish variety) and some broken dog biscuits. Always supply fresh water.

- Have woodpiles, wild corners and mounds of autumn leaves for hibernation.

- Don't rake leaves any later than the end of October or you could evict dozy hedgehogs that may then perish. NEVER rake through leaves in deep winter. Hogs hibernate from November to mid March.

- If you see a hedgehog in the day, he will be sick or injured. Place gently in a box and take to your local Wildlife Rescue Centre (see 5 The End/Useful Numbers).

- Never disturb a nest or touch young hedgehogs because the mother will abandon her young.

- Hedgehogs are wild animals and should not be kept as pets. But if you have the right garden conditions, you may be able to adopt a hedgehog. If you already have them in your garden, you can help them by building - or buying - a hedgehog house. To find out more, visit the British Hedgehog Preservation Society website at **www.softwaretechnics.com/bhps**

Assault and Battery

Bats are exquisite. Despite the Dracula tales, they don't come swooping in under cover of darkness to give you a deadly bite to the neck. Nor do they ever fly into your hair. It's an old wives' tale.

But they do eat mosquitoes and gnats and many other insects that we don't want in our gardens. In fact, one pipistrelle (Britain's commonest bat) will eat up to 3,000 insects a night. So who needs insecticides? What we need is bats!

BATS ARE UNDER THREAT. Why? Loss of habitat, roosting sites and over-use of insecticides. It's the same old story. In the last eleven years, pipistrelle numbers have dropped by 66%. Twelve of our sixteen species of bat are either Endangered or Vulnerable.

WHAT BATS NEED

- The same as all other wildlife; ponds to take insects from, trees and barns to roost in, long grass, wildflower meadows, log piles and dry stone walls, and a shady woodland edge habitat where insects congregate.
- A bat-friendly garden is a wildlife-friendly garden. If you use insecticides - don't expect bats to come calling.
- Bats eat insects. They adore moths. So provide the food they like with cottage garden flowers, a herb garden, and a good number of night-scented flowers and honeysuckle that bring the moths in. And don't forget a nice pond for the gnats and mosquitoes to whine over.
- If you want bats, try not to cut back trees or clear away dead wood. These attract insects and make good roosting places.
- Chemical treatments to roof timber are harmful to bats.
- Invite bats into your territory by installing bat bricks in new outbuildings, tunnels and walls. These offer the crevices bats require for roosting. Visit **www.norfolk-bat-group.org.uk** for more details.
- Put up a bat box. Always site near a food source (pond, woodland, mini-meadow) in a tall, mature tree. The higher the better - 5 or 6 metres from the ground is recommended.
- Try siting a few bat boxes together at different angles. In general, they should face either south-west or south-east.
- Several different types of bat box are available. Explore more by visiting **www.jacobijayne.com.**
- Remember; bats are a protected species, with fines of up to £5,000 per bat for harming them.
- Grounded bats - ring the UK Bat Helpline for instructions on how to handle and tend - 0845 1300 228.

Urban Hero

In medieval times they called him Old Reynard in grudging admiration of his native wit and cunning. The hero of countless epic tales, he always came up trumps - stealing the cheese, biting all the chickens' heads off...

We have always admired foxes. They are beautiful, clever and very adaptable. They are also one of Nature's success stories, still coming up trumps today despite the fact we hunt, shoot, poison and trap them, as well as killing 30% of them on our roads every year.

Death, for your average fox, is likely to be violent. Yet despite the high mortality - around 60% - they are surviving and doing well in the heart of our cities.

An urban fox is a joy to behold. But he has some pretty anti-social habits.

- Little is more pungent - or difficult to eradicate - than the smell of a fox. You'll know if your garden is part of his territory because scent-marking is his first line of defence. Watch out when you pick your roses - or you could be washing your hands (with lemon juice) for days.
- Digging: If he smells bonemeal in your borders, he'll think it's a food cache and dig up all the plants to get to it. Bear this habit in mind when burying the family pet. Dig deep, or the next morning you may find that poor old Fluffy has been resurrected.
- **It is strictly illegal to block an earth with foxes inside it.**
- Foxes are not vermin. They are protected from a number of deaths by law.
- Never put down poisoned food. It's barbaric, illegal and just plain foolish since you'll probably end up poisoning the neighbour's cat.
- Avoid taming foxes, or feeding them too regularly, as this could undermine their chances of survival.
- If you have an injured fox in your garden, call an expert from the National Fox Welfare Society (01933 411996) for advice on the best action to take. It is sometimes safer not to move a fox as it may lose its territory.
- If you love your garden foxes and are worried that they may have sarcoptic mange, phone 0906 272 4422 for simple home treatments.
- If you want to discourage foxes, fence in holes, avoid bone meal and use a repellent such as Renardine. Available from Grovelands on 0800 3894904.
- For more information, ring the Fox Project helpline: 0906 272 4411.

Game, Sett and Match

What a beautiful thing it is to see a clan of badgers out foraging in the hedgerow. They move across the lawn like chess pieces, their heraldic faces delicate in the dusk, snuffling and digging the soil for their favourite food. High protein, soft, easy to swallow - yes - it's worms that they live for, those long warm noodles that come stretching out of the soil with a little flick of their tails.

Except sometimes the worms don't come up that readily. Which means that the badger clan have to go on a heavy digging session, a practice that can wreak havoc with a neatly manicured lawn. Or a fine crop of vegetables. Or a golf course.

It is this habit of digging that makes them so unpopular and has brought them centuries of persecution. It has also been used as the justification for all manner of cruel sport and mad policies.

- If you have problems with badgers digging on your land, please don't take any action yourself - ring your local badger group for advice.

- Badgers are protected and badger baiting is illegal. It is an offence to kill, injure or take a badger - or damage a sett. If you have a sett on or near your land and see suspicious goings-on, call the police (999), or the RSPCA (0345 888999).
- Badgers follow traditional pathways and continue to follow them even if a road or a fence has been put across that route. Which is why you often get higher concentrations of badger road kill at specific locations.
- They will dig under or break through fences in order to follow this route. So if you are trying to protect badgers on your land from busy roads, ensure you use badger-proof fencing.
- You can reduce the road kill by campaigning for badger underpasses, particularly where new roads are being built. These are small tunnels that pass under the road at traditional pathways.
- If you have an area of road near you where badgers are regularly killed - campaign for a 'Badgers Crossing' sign to be placed ahead of the crossing place and an 'End' sign the other side so that drivers know the exact area. It sounds simple, but can make a huge difference.
- If you have badgers in your garden, try to avoid feeding them too regularly so that they become dependent on your offerings. If the ground is too hard for them to dig in summer, or too frozen in winter, you can help by offering some plain peanuts - or better still - a commercially available badger food (from CJ Wildbird Foods, see 5 The End/Useful Numbers). Never offer meat products, sweet or salty foods.
- If the earth is too dry, badgers will eat beetles, larvae, wild fruits, nuts and bulbs until the rain comes and they can get digging for worms again.
- It's not advisable to deliberately encourage badgers into your garden; remember they'll go into everyone else's garden as well, where their excavations may not be so welcome. Badgers are no respecters of prize dahlias, however splendid. So don't invite the Brocks to tea until you've checked out the neighbours.
- Injured badgers - only ever handle with gloves. If they are hurt or frightened, they'll bite. If you find one, cover it with a box or empty dustbin and anchor it with a weight. Then call the RSPCA Wildlife Field Unit (01823 480156) or Secret World Wildlife Rescue on 01278 7833250 for details of your nearest wildlife rescue. Always record exactly where you found the badger so it can be returned to its own territory after treatment.
- A note on bovine TB and badger culling policy. Is this acceptable? Visit **www.badgers.org.uk** for more information. You'll also find lists of local badger groups here.

It's a Frog's Life

Princesses kiss them. The French eat them. Prisoners are marched like them, and singers get them in their throat. Frogs are everywhere these days, except where they should be; in the countryside, at the edge of a natural pond.

With changes in farming, water pollution, pesticides, and the filling of ancient ponds to create better fields and bigger housing estates, our beloved amphibians are croaking in their thousands. Unless we wake up, these charming creatures of nursery rhyme and fairy tale will be fighting for their lives.

Frogs and toads have been around for 40-60 million years. They survived the Ice Age. But their numbers are falling fast. And they're making a last stand. In our gardens.

In the last 50 years, depending on area, around 90% of our ponds have been lost.

Just because there are lots of frogs in your pond doesn't mean they're not in trouble. The poor things are crowding into our gardens because they've nowhere in the wild to go to any more.

- Build a pond (see 4/The Action).
- A wildlife pond should not contain non-native ornamental fish (no goldfish and particularly no Koi carp). These species devour all the spawn and tadpoles - another reason for the reduction in urban frog numbers.
- Never take spawn from the wild. Ask a friend or neighbour for some of theirs.
- Don't try to separate breeding pairs, and leave young froglets well alone. They are very fragile and their skins dry out easily.

NOTE: Frog spawn is laid in clumps, toad spawn in strings, while newt eggs are laid singly within the curled up leaves of water plants.

NOTE: A newt tadpole is called an eft. It can be identified by its external gills. Toad tadpoles are blacker than frog tadpoles and often move in shoals.

- If you have thousands of taddies in a small pond and they aren't maturing quickly enough, help them along by feeding them dried daphnia (from pet shops). Make sure it's finely crumbled, though. If you go away during the feeding period, remember to drop in a pond vacation block or two, so that some food is still available to them.
- At metamorphosis time (mid-summer), frog and toadlets hang around pond edges waiting for rain. This is when they disperse. It's a dangerous time as blackbirds and other predators swoop in to feed on them, so do make sure you have plenty of ground cover around the pond.
- Keep lawns short to discourage froglets from living there. If lawns are left too long, then countless baby frogs get minced at the next mowing. Also, stamp your feet along lawn edges before strimming. This will greatly reduce frog fatalities.
- Create hibernation places with wood piles, fallen leaves, rockeries or upturned terracotta pots. Never rake leaves after end of October or you'll expose dozy amphibians.

2.The Baddies

Death on Velvet Paws

Dear Tabitha is curled up on the cushion. Ah! Ain't she cute? But what havoc, what terrible slaughter has she just been wreaking on the other side of the cat flap? Out there, she is a superbly evolved, ruthless killer. A Psychopuss.

In Britain today, there are more cats than dogs. Eight million moggies with an annual songbird kill of around 70 million. Then there's the additional death toll of about 300 million small mammals, not to mention the thousands of frogs, toads, newts, butterflies and moths that also meet their terrifying death in puss's velvet paws.

In nature, there should be one cat per 10 square kilometres. In our cities, there are roughly 200 cats per square kilometre. It's a wonder there are any birds left in the trees.

Subordinate claws

- Cats are okay - irresponsible cat owners are not. If you have a cat, please ensure it's not contributing to the wildlife carnage. Try to keep cats in between 10-11am and 4-5pm, when they are most likely to strike. To reduce the small mammal kill, try closing off the cat flap at night.
- One cat bell is not enough - any stalking moggy can silence it. Use two bells which will sound against each other, thus giving songbirds plenty of notice.
- Do consider having your cat neutered - this quells the urge to hunt and reduces the number of unwanted, hungry strays.
- Not all moggies are serial murderers, but most will kill if it's handed to them on a plate. So if you have cats, don't use a bird table, feeders or nestboxes.
- If you live in a catty neighbourhood and love birds, fence off your wildlife area with chicken wire, hurdle or prickly shrubs. Wire in gaps in hedges and try to exclude cats altogether.
- Site nesting boxes high on walls or in dense, prickly cover. Use collars of holly, brambles or wire around tree trunks and beneath nests to deter raiding cats.
- Place bird feeders, tables and baths 2-3 metres away from cover so that cats cannot approach unseen.
- The Pest Patrol; a garden gadget that emits ultrasonic blasts to deter cats and foxes (01733 315888). Cat Watch; a heat and movement-sensitive device (01763 254303). Also available from **www.conceptresearch.co.uk/index.htm**

Comrade Nutkin

Don't be taken in by the fluffy tail. Comrade Nutkin is no softy. And he has ways of getting into your peanut feeder - that new squirrel proof one you bought last week. He's across the wall in nanoseconds, abseiling down the shrubbery to your feeding station, scissoring through mesh and wire to pillage your peanuts. Comrade Nutkin is not particularly clever, but he's hugely persistent. He's also a brilliant acrobat, leaping up to six metres in one go and balancing every somersault with a twist of his tail. He bites through plastic and clears your bird table of sunflower seed in nanoseconds. Worse still, he ignores your raps on the window and runs off with a fat ball in his mouth.

Fruit and nut case. Splash out on a squirrel-proof feeder. Pricey, but well worth it to stop you going nuts.

Be cagey. Put wild bird food inside an old birdcage and stand cage on a post or bird table. Alternatively, suspend it from a tree. If the cage door is too large an opening, close this and create smaller entrances with wire cutters (remembering to smooth off sharp edges). This is an excellent way of keeping out squirrels, as well as preventing sparrow hawks from taking wild birds from the table.

Bin it. A plastic waste paper bin, bucket or plastic collar placed about halfway up the bird table pole is an excellent deterrent. Squirrels cannot grip onto the plastic and slide off.

Hot stuff. Chilli powder is said to deter squirrels because they can't bear the heat of it in their mouths. The birds don't seem to mind it, so try sprinkling some around the edges of the bird table.

On the wire. If you are feeding only seed, try placing chicken wire around the table. This will allow small birds like finches, sparrows and tits to pass in and out freely, whilst excluding squirrels, magpies and crows. Make sure it is well stapled to the wood, or a squirrel will simply tear it off with its teeth. NOTE; If larger species such as blackbirds and thrushes feed at your table, then this action is not appropriate.

Tight rope. If you hang your feeder from the washing-line, try threading two large empty plastic bottles either side of the feeder. These will spin around as the Comrade makes his way along the line so that he loses his remarkable balance and falls off.

Holed up. In spring, squirrels will bite their way into nest boxes and steal eggs and young. To prevent this, fix a 5cm square metal plate around the entry hole.

The Magpie Mafia

With his bullying tactics and swaggering demeanour, he's the mafioso of the bird world. See how he sidles along? He's up to mischief, the shifty old devil, creeping around the hedgerows looking for trouble.

The magpie is perfect for big business. A born entrepreneur. He's smartly turned-out, exceptionally intelligent, with a horrible loud voice and a taste for flash jewellery. And he is ruthless.

Rat, tat, tat, tat sounds his battle cry - just like a machine-gun - and then he pokes his cruel beak into your hawthorn bush and tears up your lovely nest of young robins.

Contrary to popular belief, magpie numbers across Britain have not greatly increased. What has changed, however, is their distribution. There are far more in our cities nowadays, because life is easier and the pickings richer. The Magpie Mafia have moved into suburbia.

Breeding pairs have a vast territory (12 acres), which means that 25%-60% of magpies in a given area don't have the space to breed. These birds will gang up and hang out in trees together like bored yobs on a street corner. Hence the counting rhyme; 'One for sorrow, two for joy...'

- Shake a large matchbox at marauders to simulate their clattering cry. Mistaking the sound for another magpie, they retreat.
- Protect nest boxes for smaller birds by surrounding them with a circle of 5cm chicken wire (see section 4/Best Nests).
- Fit a rectangle of wire above the nest. Magpies, being wary of traps, will avoid it. It also prevents dive-bombing.
- Hang little mirrors, bells, tin foil and anything that catches the light in trees. Magpies will be scared of the sudden, unpredictable movement and fly off.
- Open nests (thrush, blackbird and robin) are particularly vulnerable. If you have such nests in your garden, you can keep an ear out for magpies. Whenever you hear that 'rat a tat tat,' you know it's time to chase them off.
- Make sure your hedges are impenetrable, tightly planted with prickly species, and with no big holes and patches that magpies and jays can get through.
- Magpies, sparrow hawks, rooks and crows observe nesting behaviour from high above. If they see a lot of movement, they'll know there's a nest. So, the more cover a bird has, the safer it will be. Protect your songbirds by providing shrubs and trees, so that birds can hop from hedge to branch undetected. Good ground cover and tall plants in borders also obscures movement.

The Plot

Chemical Warfare

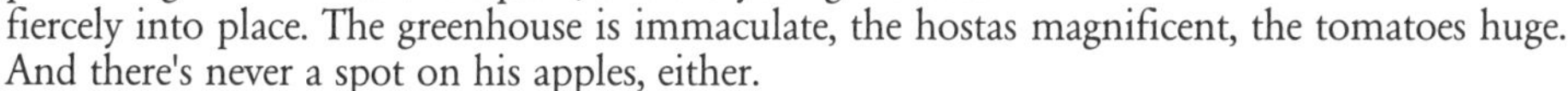

You can *always* recognise the orthodox gardener.
His lupins stand to attention. His dahlias salute as you go past. His garden is neat and square, and everything is staked fiercely into place. The greenhouse is immaculate, the hostas magnificent, the tomatoes huge. And there's never a spot on his apples, either.

To achieve horticultural splendour, he poisons the earth - and the wildlife that depend on it. Weed-killers, insecticides, slug pellets, chemical fertilisers, moss killer, lawn feeds and path cleaners.

But remember - he shares your wildlife. Your robin is his robin, your hedgehog his. And the environment is the responsibility of us all.

DID YOU KNOW?

The average garden receives a heavier input of pesticides than an intensively managed farm field.

One tablespoonful of garden insecticide poured down the drain, will pollute the water supply of 200,000 people for a day. Imagine what it does for wildlife.

STOP THE CHEMICAL CARNAGE NOW!

- Never spray fungicides or herbicides. Always weed by hand. To limit weed seeding and spreading, avoid formal planting design (lots of bare earth) and increase ground cover.

- Don't use bleach on paths and patios. It's vicious stuff. Detergent solution is appalling, too. Use garden hose to jet-spray aphids off your roses. Or wipe them off with your fingers.

- Avoid insecticides altogether. Nature has provided her own fly control in the form of ladybirds, beetles, spiders, hoverflies, lacewings, etc. So why destroy this natural flypaper with chemicals? Not to mention all the butterflies and bees.

- Slug pellets? No way. Widespread killers of all that we wildlife gardeners hold dear. Find the information you need and many alternatives below (The Slug & Pellet).

- Ant-killer doesn't just kill the ants. It contains **bendiocarb**, which kills the birds as well. Boiling water poured directly into the nest is the safest non-chemical way to deal with infestation.
- Please avoid puffing ant powder all over the place. Instead, if you have to, try using a gel-based product - a tiny blob at the entrance to the nest will be enough, but do remember to cover with a plant pot to stop birds and other insects having a nibble.

FIND ANOTHER WAY

- Don't plant all your roses together. It's rather like laying on a buffet for aphids.
- Confuse greenfly by mixed planting - intersperse roses with plants they hate (sage, rosemary, lavender, borage), so it's harder for them to home in. Herbs also attract the goody insects that feed on aphids.
- Cow parsley, sweet ciceley and fennel attract lacewings. A lacewing larva eats more than 100 aphids an hour. Ladybirds are equally voracious. Entice them into your garden with golden rod and prince's feather. And remember; hoverflies are goodies, too.
- 98% of lowland raised bogs have been destroyed through removal of peat for potting compost. This delicate habitat cannot be replaced. So please use peat-free compost whenever you can. There are plenty of good substitutes on the market. All you have to do is ask.
- Avoid all rose, lawn and plant feeds that are not organic. Chemicals induce a false surge of growth that is neither healthy to the plant nor the environment. Fish, blood and bone is an excellent fertiliser, and so is good old manure. Both are totally natural and (as long as they have no chemical additives) organic.
- Everything's a balance and weeds can invade gentler native species and quickly take over the wildlife garden. For really stubborn weeds like thistle, dandelion, bindweed and ground elder, you may need to resort to a glyphosate herbicide, but only in exceptional circumstances. Visit the Pesticide Action Network website at **www.pan-uk.org** for their recommendations.

IMPORTANT NOTE

In autumn 2003, a number of pesticides and weed killers - of which 81 were for garden use - were banned. It is now illegal to use or even store them. To check out the list of banned products, visit the Pesticide Action Network website.

If you have any garden chemicals you need to dispose of, DO NOT THROW IN THE BIN OR DOWN THE DRAIN. Ring your local council. They will inform you of the appropriate action to take.

The Slug & Pellet

Not a pub

When God created the SLUG, he wasn't considering the hosta, the lupin and the delphinium. He was merely thinking about the provision of food for other, more attractive creatures. Because what are slugs really, but a meal on the move? And while they move, they eat. Gliding across your lawn on their horrible orange foot, feeling the air with tender horns, their heads all pointing in the same direction as if they're going to church. *Except that their chosen place of worship is your lupin bed.*

Worse still, slugs don't even need a mate to reproduce - it's a DIY job, ending in the laying of thousands of eggs. After which they slither off and leave YOU to feed their babies just as you once fed them - with your prize petunias.

Let them eat pellets you scream, scattering handfuls around your flowerbeds.

STOP!

Slug pellets contain a deadly substance called **metaldehyde**. This is highly toxic to garden creatures, pets and aquatic life. So when Fred Slug nibbles a pellet, he immediately tries to purge himself of the poison by producing vast amounts of mucus.

Since a slug can't move without slime, he is effectively immobilised. Left in your flower-bed, slowly desiccating to death in the full glare of the sun.
Unless, of course, it happens to rain.

Once wet, Fred can quickly re-hydrate and slither off to live another day - or even two - with traces of poison still festering away inside him. He looks healthy enough to a passing hedgehog, who gobbles him up, and...you can guess the rest.

Managing without metaldehyde

- Put a 5cm band of copper around plant pots or bender boards. It gives slugs and snails a nasty shock so they won't cross it.
- Grow slug-attractive plants (eg. hostas, lupins), in tubs until they are big enough to survive attack. Ensure tubs are slug-free, then put a thick 2.5cm band of vaseline around the outside of the tub.
- Try mixed planting - surround delphiniums, etc with foxgloves, geraniums, nigella, garlic, sage, parsley, rosemary or nasturtium to deter slime travellers.
- Gastropods love comfrey. You could try growing some near your hostas in the hope of distracting them. You could also put rings of cut comfrey around vulnerable plants.
- Collars of holly and other pricklies might deter slugs and snails. So might crushed egg shells and sand. So might bits of broken china. But don't hold your breath.
- Grow all seed in trays and pots, rather than sow directly outdoors. Surround trays with a thick barrier of fine sharp sand or grit.
- In winter, turn soil regularly to expose hidden eggs which then perish on the surface. Cut vegetation back. Improvise beer traps out of disposable plastic cups half-filled with beer and then sink into the earth. Always put in an exit leaf so that beetles and caterpillars can climb out.
- Invite slug predators into your garden by building ponds and hedgehog houses. Entice blackbirds, thrushes, robins, starlings, etc with scraps of cheese and apples. It takes about three years to get the balance, but it **really does work**.
- Try hand collecting slugs (on wet nights with a torch). Feed them to the local ducks. Drown them in a bucket of water. Stomp on them. Or snip them in half with garden shears - eek!
- Better still, allow that gastropods were created for a reason. They're part of a far wider plan than we'll ever see. So why not accept them and live in peace.

Things with Wings

Sadly, for the last few decades, things with wings have been systematically slaughtered with insecticides. In the past, we just didn't know enough about the products we were using. That pesticides would kill the goodies as well as the baddies.

Add to that modern farming methods, aggressive weed control and new housing developments encroaching on the countryside, and you have a clear picture as to how 97% of our wildflower meadows - a crucial butterfly habitat - have been destroyed.

Flydown. Insecticides kill butterflies. Herbicides kill the plants that butterflies lay their eggs on. So it's destruction from all angles.

Fatal statistic. Two-thirds of British woodland butterflies are under threat. The decline of some species is at 94%. Moths are also disappearing at an alarming rate

Turnaround. Do your bit to help butterflies and moths. Create a butterfly garden. It's simple to do and will also attract other benign insects, including bees, ladybirds, lacewings, hoverflies, etc.

To bee or not to bee. Although a butterfly garden will attract bees, there are some species that are bee-specific. Bees love bell and tube-shaped flowers. They love heathers in Spring, sedum and ivy flowers in autumn, and they adore clover. So let some grow in your lawn.

Bee friendly. Cottage garden flowers are favourites, with foxgloves, hollyhocks and penstemons topping the list. Comfrey and borage are excellent, too, and as many flowering herbs as you can fit in. Honeysuckle, catnip and the humble daisy are also popular.

Detox. Say no to toxic chemicals (see 3/Chemical Warfare for alternatives).

- For more information visit the Butterfly Conservation website at; **www.buttefly-conservation.org**.
- For bee boxes, bug and ladybird boxes, contact the Oxford Bee Company at **www.oxbeeco.com.**

BUTTERFLY GARDEN

Sun lover. Butterflies love to bask in warm sunshine, so choose a really sunny corner for your butterfly garden. Make sure it's sheltered, because they don't like wind.

Hot spot. Butterflies spend a lot of time basking and warming their wings. So create the chill-out zones they love; nice flat rocks, rotting logs. And don't forget a sandy puddle for much-needed moisture.

Landing pad. Butterflies are short-sighted and rely on smell to locate flowers. They are attracted by masses of flowers of the same type grouped together. They also like large flat-topped flowers so that they can land easily.

Nectar card. Plant up with butterfly and moth-attracting species. Remember that you need both nectar-producing flowers and the less attractive larval food plants for the caterpillars. It is no good having one without the other.

Butterfly flowers:

Asters, Chives, Phlox, Candytuft, Alyssum, Zinnia, Heliotrope, Buddleia, Marigold, Lantana, Sedum, Knapweed, Milkweeds, Valerian, Yarrow

And don't forget honeysuckle and night-scented stocks for the moths.

Caterpillar plants:

Dill, Broccoli, Parsley, Carrots, Fennel, Hollyhock, Cabbage, Willow, Elm, Wild cherry, Privet, Lilac, Nettles, Docks

Plus others specific to individual butterflies and moths.

Safe Seccateurs

Spring is sprung and the lovely green season is upon us. Cast your eye over the garden; the birds are hopping around gathering twigs, and the frogs are mating and churning out spawn by the bucket load. Leaves are sprouting, and all of a sudden the hedge starts to look a bit messy...

WAIT!

Before you head to the shed for the seccateurs, remember that spring is NOT the time for cutting back. It's the time for new life. And if you start lopping off branches and trimming hedges, you'll disturb birds and destroy nests.

DO THINGS AT THE RIGHT TIME

SPRING

Watch the birdie. Every abandoned nest means an average of three fewer birds in your neighbourhood this year. So please don't make major garden alterations until the end of July when nesting is finished. Even hedge trimming is not recommended anywhere near a nest, so always check for nests before you go for the chop.

Stick it. Birds, when flying at top speed, cannot tell the difference between real and reflected images and so they just carry on flying into what they see as sky - and break their necks. If this is happening with any of your windows, you can prevent it by simply placing a net curtain across the offending glass. The alternative is to make - or buy from the RSPB - a large black silhouette of a flying bird (up to half a metre across the wings) and stick that in the centre of the window.

The Frog Mince. Your garden has been quiet and wildlife-friendly for months throughout late autumn, winter and early spring. But suddenly it is filled with whirring blades and cutters. Animals don't know this. So please check lawn edges for frogs and hedgehogs before starting up the strimmer.

Ponder this. In *spring,* clear ponds of dead leaves and blanket weed **before** spawn hatches, but do check first for overwintering newtlets.

AUTUMN

Ponder that. In *autumn,* fish out fallen leaves but do not dig around the bottom of the pond. If you're using net over the pond to catch leaves, use covered wire rather than light plastic netting and raise it 20cm from the ground so animals don't get trapped beneath.

No cut backs. The correct time of year for cutting back is autumn BEFORE hibernation. It is better for wildlife and for trees. Avoid cutting back evergreens such as holly and laurel in deep winter as wild animals and birds are very dependent on these for shelter. Once the other trees are bare, this is their only protection from snow, wind and miserable weather.

Go to seed. Try not to tidy everything up in autumn. Leave seed heads on plants and shrubs for winter food. And give your windfalls to the birds.

Pot roast. Never pile up vegetation for autumn bonfires over a period of weeks, since it will quickly become a refuge for wildlife. So please poke through unlit bonfires before setting fire to them. And after a day's gardening, do check inside sheds and outhouses before closing up, in case any creature has become trapped inside.

Box it up. Put up nest boxes in October. Birds need at least six months to get used to a new box. When fitting, try to remember what your chosen site looks like in spring when the trees are bare. Is it under good cover? Is it high enough? Is it predator-proof?

Be swift. If you are lucky enough to have swifts or swallows nesting on your premises, count yourself blessed. Please don't knock their nests off the wall for the sake of a few droppings. These birds fly thousands of miles to rear their young on your wall. Their time here is short. So cherish them.

Log on. Hibernation time is nigh. Make piles of logs and stones in sheltered corners. Avoid raking. Shovel fallen leaves into drifts. After October, don't disturb dead leaves.

The Action

Best Nests

Spring is **nesting** time and that means being careful in the garden and keeping an eye out for predators. At this time of year your garden birds need a helping hand, especially those that nest in the open rather than in boxes - thrushes and the blackbirds, for example. Somewhere around 90% of our native blackbirds' attempts at nesting end in failure. And we've all witnessed the destruction of nests by predators, despite the valiant efforts of distraught parent birds. It's a heartbreaking sight.

GIVE YOUR BIRDS A HELPING HAND

Nesting instinct. Ensure nests are as safe from garden predators as possible. If you have hedges or other sites in your garden that birds regularly use, you can help protect them from aerial attack by magpies, jays and crows.

Wired up. A large square of chicken wire over the top of the hedge where birds are nesting will put off destructive birds, while filling in gaps and spaces within and below patchy hedges will help deter cats. This can be done with chicken wire or planting. If you choose to use chicken wire, ensure that the birds themselves can get in and out of the hedge safely and will not get trapped.

Hedge your bets. Do not surround the whole hedge with wire - just the top directly over the nesting area. The reason for this is that birds of the crow family and particularly magpies are very wary of wire because they fear traps. Also build in protection for the future by planting prickly species like hawthorn, holly, blackthorn, and lovely dog rose into your hedge. This will improve nesting, boost bird numbers, and give a single-species hedge a really beautiful and natural makeover for the future.

Gift box. Not only do nest boxes provide protection from predators, they also give warmth, and shelter from wind and rain. In addition, birds may use them for winter roosting in harsh weather. But remember, bird boxes come in all shapes and sizes and some are more decorative than practical. Anything plastic or shaped like a little cottage with chimes and Home Sweet Home signs, is more likely to appeal to a human than a bird, and probably only last about five minutes outside. So stick to the approved makes and don't pay over the odds. Let's support wildlife, not the gift industry.

Mix and match. Find out which boxes are for which type of birds and match your box to your species. Boxes can be enclosed or open-fronted, and hole and size dimensions vary with each bird. So identify your birds and then give them a choice. Rarer bird boxes are available as well as the standard ones, and blackbirds and robins will sometimes nest in an open-fronted box, so it is worth putting one out. Sparrow colony boxes are a must (available from RSPB).

In-site. Position is everything, and should mimic the nesting habits of the bird in question as closely as possible. Always choose a cool, shaded spot away from direct sun, and avoid bare south facing walls, or you'll cook your chicks!

- Ensure there is good cover, that the box is at the correct height, is protected as well as possible from predators, and secure enough not to be blown off the tree in a high wind.
- Avoid windy corners and never position entrance hole in a draught. Never site a nest box over or alongside a road - or your nestlings will become a statistic.
- Nest boxes should be in place by early March at the latest, but are more likely to have visitors if you get them in place the preceding October.

Top of the box. The best boxes are set high up on tree trunks away from the predations of cats, or in or behind thick shrubby cover, ivy or clematis. You can stop cats climbing up trees by placing a collar of barbed wire high up the tree and underneath the box.

It's a wrap. Magpies, cats and squirrels will all target nest boxes so make sure yours is safe. You can protect nest boxes with a 'ball' of 5cm chicken wire arranged 30cm above and around box. This will deter predatory birds and squirrels. Make sure you fix or support the wire so that squirrels can't squash it down.

Put a lid on it. A perspex hood or large dish cover, tray, or saucepan lid fitted over a nest box will provide good protection from swooping crows, sparrow hawks and other predators.

Guardian angel. To stop weasels, squirrels, woodpeckers, cats and magpies from clawing at boxes and stealing nestlings, fix a special Nest Box or Tunnel Guardian to the front of your nesting box. These are short plastic tubes with ladder attached so birds can come and go safely. Available from CJ Wildbird Foods via the RSPB at **www.rspb.org.uk**

Flight of Fancy. Fully feathered "abandoned" fledglings are never abandoned. And never lost. They have usually had a stab at flying, under parental supervision, and there will always be an adult bird close by.

- Avoid the temptation to touch or move fledglings - if they become separated from the parent they will die.
- If you think the bird is in danger, gently move the young bird away from predators (or traffic) to a place of safety nearby, and watch from a distance. You'll soon see the parent approach.

NB. Never put fledglings in containers or boxes, unless they are injured or being transported. Also, never put them in garages, greenhouses, or take them indoors.

Fit Birds

It costs a bit more than tuppence a bag these days but it's worth every penny.

Looking after birds isn't just about feeding them, though. It's about taking responsibility for their survival; protecting them from bad weather, predators, pesticides and hunger.

GOOD-LOOKING BIRDS

- It doesn't take much. A handful of sunflower seeds, a few crumbs of cheese. The RSPB recommends year-round feeding to improve general and nestling survival.
- Feeding stations should be placed in the open, away from shrubs, with a clear view of approaching predators.
- Avoid positioning bird tables too close to windows, as they can easily fly into the glass and break their necks. It's nice to watch them up close, but not *that* close.
- Always provide water - a birdbath is essential. Plinth-mounted birdbaths are the safest, but an upturned dustbin lid on bricks will do. Or, failing that, a shallow dish. Keep water clean and chemical-free. Break ice in winter, or thaw with hot water poured over the ice. A small aquarium thermostat can be rigged up to keep your bird bath thawed. **Never** use **any** additives, particularly salt or anti-freeze.
- Some songbirds are seed-eaters, others eat fruit and berries, worms and insects. Try to supply these natural foodstuffs with shrubs such as hawthorn, blackthorn and alder.

Berries that pull the birds. Holly, rowan, cotoneaster, hawthorn, elder, apple, crab apple, dogwood, rose hips, teasels, wild strawberries, redcurrants, blackcurrants, raspberries, blackberries, gooseberries and Japanese quince. These will encourage thrushes, blackbirds and fieldfares as well as the usual pigeon brigade and the rarer waxwings and warblers.

Seedy characters. Resist the temptation to dead-head all your flowers. Let some go to seed for the finches and sparrows. Try growing wildflowers that attract birds and insects as well as those that produce good seed. Seed-bearing trees are also good – native species being the very best you can offer native birds!

Attractive species. Birch, sycamore, willow, ash, St John's wort, crocus, teasel, fat hen, dock, poppy, field forget-me-not, evening primrose, cow parsley, yarrow and meadow sweet. In fact, any wild flowers are of use, either because they attract flies and insects or because they make prolific seed for sparrows and finches.

Be hospitable. Have a weedy patch in your garden. Docks, nettles, thistles, plantain and some grasses attract butterflies and moths, providing birds with a protein-rich supply of caterpillars. Nasturtiums are also good. Not to mention cabbages! And plants like teasels, evening primrose and red campion are popular with small flies on which birds feed.

Log on. Dry stone walls and compost heaps are good sources of grubs and beetles. Ponds also attract insects on which birds feed.

Feed the Birds

Going nuts. Unless safely housed in a wire feeder, peanuts can easily choke young birds. Any loose peanuts should either be grated, or picked out and thrown in the bin. Substandard nuts have sometimes been found to contain aflatoxin, a poison that kills birds, so make sure yours are fresh and come from a reliable source.

Diet of Worms. Remember that blackbirds, robins and thrushes, etc are NOT seed-eaters, so put out crumbled cheese, little bits of bread rolled in unsalted dripping, and any worms and leatherjackets you happen to dig up. Avoid putting out hard, dry crusts as these can cause choking in smaller birds. NOTE: In dry weather, water hard earth to bring worms to the surface for birds.

Get soaked. Always soak dried foods before offering to

the birds. Desiccated fruits such as sultanas, raisins, etc, are so dry they swell up inside the bird and can dehydrate them. Fresh fruit is always better.

The Big Apple. Thrushes, blackbirds and fieldfares ***adore*** apples. In autumn they will feast on windfalls, but the rest of the year (particularly through winter) they have no access to this brilliant source of energy and nutrition.

An apple a day. You can conserve apples by wrapping them in newspaper and storing in a cool, dry place, and then offer them through the winter .In the bitter months of January and February, treat your garden birds to some bought apples - stockpile them when they're on offer and put two or three on the ground every day, particularly in snow or frosty weather. They'll be skinned by lunchtime!

Chopaholic. Apples should not be chopped as they are harder to eat and can cause choking. Mother Nature offers her apples whole. That's the way the birds like them.

Sunflower power. All the seed-eaters love them, and a feeder placed in an open space but near cover, will bring in hordes of birds; chaffinches, bullfinches, green finches, goldfinches, siskins, tits, etc. Other species will come to collect seeds dropped from the feeder; pigeons, pheasants, sparrows, dunnocks. It'll be a real party out there!

Fat of the land. Fat balls, half coconut shells (fresh, *never* desiccated), and whole peanuts in their shells can be threaded with strong nylon twine and suspended from bird tables and trees. Make sure you site them safely out of the way of cats and ensure the peanuts are well secured. Unfortunately peanuts generally fall prey to grey squirrels, so it is probably worth investing in a squirrel-proof feeder.

Feeding frenzy. Feed wild birds regularly right through the year to increase the chances of nestling survival. And try to be consistent with your feeding - don't have a craze and then suddenly lose interest. Remember that additional chicks will have been reared on the basis of food availability and these birds will be dependent on your offerings. If you go away, see if you can find someone to continue feeding until you return.

NB. Keep feeders, bird baths and tables scrubbed clean (no chemicals), and never leave mouldy food lying around. Wild birds are susceptible to salmonella and other diseases.

Unwanted guests. if you are feeding the birds, you can expect an occasional visit from less welcome creatures. One of these might well be Roland Rat. If so, site feeding station away from house and try using bird feeders instead of a bird table. Avoid scattering grain or giving scraps. If you are visited regularly, stop feeding altogether, and Roland will stop calling. If rats are coming indoors, ring your local city council and ask for Environmental Services. Never put down poison yourself.

Supergrass

The Garden Meadow

For those of us who remember acres of beautiful, butterfly-filled meadow, it comes as something of a shock to learn that 97% of our wildflower meadows have now gone. This may bring all manner of insects and creatures to the brink of extinction.

The good news is that we can help. Even in small gardens we can reconstruct mini-meadows to help provide the food and habitat our butterflies so desperately need.

In fact, leaving part of the lawn to grow wild is not only a guaranteed way of bringing in wildlife, it also makes a striking feature. The contrasts of colour and texture are sensational. Not to mention all the different shades of green.

- All you have to do is choose an area of the lawn where you'd like to have long grass. Think of a shape (circle, semi-circle, square, etc) and mark it out with chalk or string.

- Then, when you next get out the lawn mower, leave this area uncut.

Introducing wildflowers into grass to create a meadow is no easy task and wildflower germination is erratic. So here is a fool-proof method;

- Perennials are easier to establish than annuals, but remember they won't flower the same year you sow them.

- To guarantee the best yield, sow wildflower seed into pots (not trays) from September to October. Sprinkle a few seeds in each pot, from a mix of oxeye daisy, buttercup, selfheal, common vetch, yellow rattle, ribwort plantain, viper's bugloss. And don't forget your annuals; corncockle, chamomile, field poppy and cornflower.

- Over-winter outside, protecting seedlings from slugs (see 3/The Plot).

- Once each plant is strong and well-established, plant out perennials in July/August in the position you want them to flower the following year. Dot them through the grassy area.

- Keep building stock in this way until you have a small meadow that will flower from the end of April through to the beginning of July.

- Cut area in mid-July, after flowering, and leave the mowings where they fall for three days so that wildflower seed can be set for the following year. Then rake off and compost the hay.

- Keep grass at around 8-10 cm in length thereafter.

The Trellis Tunnel

Trellis fixed to walls and then planted with climbers such as rose, ivy and honeysuckle will make an excellent refuge for birds. But a trellis tunnel is even better.

For a start it is three-dimensional. And for birds it's the nearest thing to a thicket - even better, in fact, because it has height as well as depth.

It's also a great way to conceal unsightly outbuildings, garages and sheds.

The trellis tunnel can either be free-standing or used with a wall. If you are able to incorporate a wall, this is better, as it provides stronger support, and has more surfaces adjoining it (other walls, roofs, etc) to create a multi-dimensional effect.

Within two years the climbers will have grown across the batons and all over the wall behind, creating a living tunnel of dense vegetation. A wildlife magnet!

DIRECTIONS

- Choose your site. Make sure it is in a remote a part of the garden away from paths, doorways and roads.
- Erect a trellis fence in a design of your choice - about 2 metres in front of the building you wish to conceal, and parallel to it. Fit wooden batons from the top of the trellis to the wall. These can slope upwards to the roof.
- If your tunnel is to be free-standing, erect two trellis fences parallel to each other, and join them across the top with either batons or hoops of wire.

Then plant up with a mixture of the following;

Honeysuckle
Climbing rose
Sweet briar
Ivy
Clematis
Wild sweet pea
Vetch
Climbing hydrangea
Passion flower
Jasmine (if south facing)

For more depth, sow wildflowers to the front of the trellis to give it a hedgerow effect.

The Humble Hedge

In the last fifty years over 200,000 miles of hedges have been destroyed. This is disastrous for wildlife. Hedges are green corridors across wide spaces, offering wildlife shelter and protection from predators. They also attract insects and other creatures on which wildlife feed, as well as providing wildflowers for butterflies and bees. And of course, they are vital for nesting and roosting.

Hedges are the arteries of the natural world, connecting different parts of the habitat on which wild creatures are so dependent.

And there just aren't enough of them left. So forget the fencing. Build a hedge.

- If you already have a hedge, but, like many garden hedges, it is a single species variety, then think about introducing some other species into it. Not only will it add depth and variety to an existing hedge, it will attract a far greater selection of wildlife. And don't forget to plant wildflowers around the base and in front of your hedge.
- A real hedge is natural and has evolved over a long period of time. It is only cut back once a year in winter and otherwise pretty much left alone. So forget perfect lines of uniform clipped green. Nature's hedges are soft, curving, dappled, intricate and lovely. How can you bear to be without one?
- There are two main species in a native hedge - **hawthorn** and **blackthorn** - with others mixed in. The combination of these will depend on locality and soil type.
- Other species may include; field maple, hazel, common alder, common dogwood, wild privet, crab apple, alder blackthorn, field rose, dog rose, guelder rose, sweet briar.
- Wildflowers for hedgerows; red and white campion, tufted vetch, viper's bugloss, self-heal, ox-eye daisy, knapweed, cow parsley, yarrow, hedge woundwort, self-heal, wild basil.
- To order large quantities of wildflowers for hedgerows, visit **www.emorsgate-seeds.co.uk** or ring John Chambers Wildflower seeds 01933 652 562. For smaller quantities, contact Landlife Wildflowers at **www.wildflower.org.uk.**
- For more about hedging visit the National Hedge Laying Society at **www.hedgelaying.org.uk**.
- The British Trust for Conservation has produced an excellent book; Hedging (£13.95). For more details visit their online bookshop at **www.shop.bctv.org.uk/shop/index.**

FOND OF PONDS

Instructions for Building a Wildlife Pond

Ponds are an essential part of the wildlife garden. They create refuges for wildlife, provide food and insects for birds, not to mention drinking and bathing areas. They bring a whole new focus to the garden and are wonderful to watch.

Spring is the best time to build a pond.

- Choose a level site. Try to incorporate a marshy area if you have one, which can then take run-off from the pond and be home to native pond-edge plants.
- Position - some form of shelter - a wall or small bank, offers protection against frost and wind, but is not essential. Avoid putting ponds under trees or in full sun. The best sites are those that get sun for about half the day.
- Depending on overall size you choose, the centre of the pond should not be deeper than 80cm, becoming shallower towards the edges (to allow for spawn-laying).
- Don't forget to have a sloping exit on one side (for small mammals and invertebrates that might fall in), and a plant shelf (about 25 cm wide and around 20-30 cm below water level).
- Mark out the shape using string or chalk.
- Once you have dug out the required shape, line with a layer of sand. Then cover this with an old piece of carpet or a thick layer of newspapers to prevent liner damage.
- Use a butyl rubber liner. Always get a bigger size liner than the measurements of the hole to allow for water, stones and planting.
- Work out required liner size by calculating;

    1. How deep your pond should be (no more than 80cm at its deepest point even for a large-sized pond, with varying depths).

    2. The length of liner required - it should be the overall length of the pond plus twice the maximum depth.

    3. The width of liner required - overall width of pond plus twice maximum depth. This will allow for variations in depth and sloping sides.

- Never cut liner to final size until pond is filled and edged.

- Once liner is in place, cover it with a sand mix. Then fill pond with water.
- Edge with stones - preferably rockery-style for a more natural look, or dry stone wall to give cover and hibernation places.
- Ensure there is always plenty of ground cover for creatures to move in and out of the pond safely. Long grasses and thick clumps of plants provide this.
- Resist the temptation to stock with goldfish - or anything much. Be patient and wait for the wildlife to move in. It won't be long until the first water boatman appears. Little water snails seem to come from nowhere. Dragonflies investigate. Gnats and other insects bring in the birds...
- To start things off more quickly, add a bucketful of water from a neighbouring pond. Introduce frog and toad spawn in spring - but never from the wild. Just ask a friend for some of theirs.
- Plant with only native species - non-natives pose a great threat to our rivers and wild waterways.

Some native species;

Yellow flag iris
Flowering rush
Marsh marigold
Greater spearwort
Water crowfoot
Water mint

NOTE. Safety precautions are essential if there are children anywhere in the vicinity. Ponds should never be too deep, and either fenced off or covered over with weld mesh.

Bubble Pools

If, due to lack of space or very small children, you are unable to have a wildlife pond, do not despair. You can still attract birds and insects into your garden or patio with a bubble pool. A half-barrel, tub, old sink or bath will do. Simply put in some pebbles, then a small pump, and then cover this with more pebbles so that the water bubbles merrily over them. To add more interest for wildlife, you could put in some native water plants or rushes. Never use chemicals in ponds or bubble pools.

The End

Wildlife Garden Checklist

- Dry stone walls
- Log piles
- Mixed-species hedges with hedgerow wildflowers
- Shrubs and trees - all native and preferably berry-bearing - with some evergreen for winter shelter
- Trellis - arches and tunnels
- Pond and/or bubble pool
- Bird table, feeder and birdbath
- Compost heap
- Water butt
- Let plants go to seed
- Climbing rose, clematis and ivy
- Mini meadow
- Butterfly and bee garden
- Wild rose and honeysuckle to colour up the hedges
- Patio tubs full of wildflower annuals
- Hedgehog house, nest boxes, etc
- No pebbles, decking, concrete or all-over paving
- No garden chemicals
- Peat-free potting compost
- Organic fertilisers
- No bonfires
- Garden in keeping with the wildlife year

A Few Further Ideas

- Grow annual seed in cracks of patio, especially field forget-me-not, camomile and poppy.
- Raise odd patio slabs and plant the spaces with butterfly and bee attracting wild flowers.
- Plant wild snowdrops and wild crocuses in the lawn directly in front of the meadow area.
- Try inter-planting the herbaceous border with wild flowers.
- Create a wild flower rockery, with plenty of stones for hibernating creatures.
- Make a wild area at the back of the garden, perhaps near the trellis tunnel, where you can plant the taller species such as wild comfrey, teasel, cow parsley and mullein. Interplant with field buttercups, red campion and oxeye daisies.
- Buy attractive tubs and then plant them up with nettles and docks. This will contain root spread but still provide butterflies and moths with somewhere to lay their eggs. You can always plant in some wildflowers as well to make the tubs more attractive.
- Grow cabbage and broccoli alongside the tubs to create a caterpillar corner.
- Have a soft fruit corner in your garden - raspberries, blackberries, gooseberries, redcurrants, etc - and leave some for the birds.
- Plant an apple tree.

In the Home

Here are a few simple ways in which you can extend your care of wildlife into the wider environment.

- Use eco-friendly bio-degradable detergents and washing powders.
- Try to reduce the amount of these products you use. One squirt will do the job. And you only need a couple of drops of floor cleaner in the bucket - not half the bottle.

- Avoid bleach.
- Wash more by hand. Half-empty dishwashers and washing machines are a waste of water.
- Use half load and short programme settings whenever you can.
- Don't leave taps running. And turn off the tap when you're brushing your teeth.
- Take more showers and fewer baths.
- Water gardens with a hand-held hose.
- Use less heating. Wear more jumpers. Insulate lofts and roof spaces.
- Turn lights off.
- Avoid using aerosols, cling film and plastic carrier bags.
- Kettles; save energy by only boiling the amount of water you need.
- Try not to use disposable nappies. Traditional terry nappies are healthier, less expensive, and far greener.
- Recycle everything you can.
- Walk to work, share lifts or take the bus.
- Eat as much organic produce as you can – or grow your own.
- Avoid GM products.
- Be careful what you put down the sink - particularly chemicals, antibiotics, hormone supplements and other unwanted medications.
- Always pick up litter. Not only is it unsightly, it is a hazard to wildlife. Plastic bags cause suffocation, broken glass and empty tins cause cuts and lacerations, and twine and thread get caught around feet and tighten, causing gangrene.

Useful Addresses

Cat/Fox/Mole Deterrents;
CATWATCH Concept Research UK
37 Upper King Street
Royston
SG8 9AZ
01763 254303
www.conceptresearch.co.uk/index.htm

CJ Wild Bird Foods
The Rea
Upton Magna
Shrewsbury SY4 4UB
0800 731 2820
www.birdfood.co.uk

Ernest Charles (Wild Bird Food)
Freepost
Crediton
Devon EX17 2YZ
0800 731 6770
www.ernest-charles.com

The Soil Association
Walnut Tree Manor
Haughley
Stowmarket
Suffolk IP14 3RS
0117 314 5000
www.soilassociation.org

Pesticide Action Network
Eurolink Centre Unit 16
49 Effra Road
London SW2 1B7
0207 065 0907
www.pan-org.uk

RSPB
The Lodge
Sandy
Bedfordshire
Herts SG19 2DL
01767 680551
www.rspb.org.uk

The Wildlife Trusts
The Kiln, Waterside,
Mather Road, Newark,
Notts NG24 1WT
01636 677711
www.wildlifetrust.org.uk

The Woodland Trust
Autumn Park
Grantham, Lincs
NG31 6LL
01476 590808
www.woodland-trust.org.uk

Landlife
National Wildflower Centre
Court Hey Park
Liverpool
L16 3NA
0151 737 1819
www.wildflower.org.uk

Useful Websites and Contacts

Badger conservation;	**www.badgers.org.uk** **www.mammal.org.uk/badger.htm**
Bat conservation trust;	**www.bats.org.uk**
Bat boxes;	**www.jacobijayne.com**
Bat bricks;	**www.norfolk-bat-group.org.uk**
Bee boxes, etc;	**www.oxbeeco.com**
British Dragonfly Society;	**www.dragonflysoc.org.uk**
British Hedgehog Preservation Society	**www.software-technics.com/bhps**
British Trust for Conservation;	**www.bctv.org.uk**
British Trust for Ornithology;	**www.bto.org**
Butterfly Conservation;	**www.butterfly-conservation.org**
English Nature;	**www.englishnature.org.uk**
Fox Project;	**www.foxproject.org.uk**
Fox Welfare Society;	**www.nfws.org.uk**
Froglife;	**www.froglife.org**
Green Queen	**www.greenqueen.co.uk**
Hedging;	**www.hedgelaying.org.uk**
Herpetological Conservation Trust;	**www.herpconstruct.org.uk**
Mammal Society;	**www.abdn.ac.uk/mammal**
Plantlife;	**www.plantlife.org.uk**

Wildflowers in large quantities; **www.emorsgate-seeds.co.uk**

Smaller quantities;
Landlife Wildflowers **www.wildflower.org.uk**

Native wild plants; **www.floralocale.org**

Injured/Abandoned Wildlife

RSPCA Wildlife Field Unit **www.rspca.org.uk**
Telephone; 01823 480156

Secret World Wildlife Rescue **www.secretworld.org**
Telephone; 01278 783 250

Wildlife Helpline National Service **www.wildlifehelpline.org.uk**
Telephone; 01522 544 245